101 Questions for
The End of the World

Coffee Table Philosophy

Volume 10

J Edward Neill

Tessera Guild Publishing

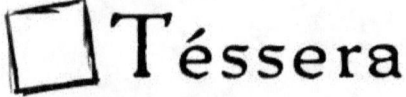Téssera

ISBN-13: 978-1532962165
ISBN-10: 1532962169

Commentary

A part of me is ever optimistic.

Another part of me…not so much.

I've always hoped for more out of the universe. And what I mean by more is that I want a greater meaning. A deeper understanding. The ability to one day, even if after my own death, to pull back the curtain and see what's really going on.

I want to know.

Let's clarify.

I want to know *everything*.

So do you.

Probably.

And that's the problem, isn't it? For though we may spend our entire lives learning, relearning, and discovering, we'll never know it all. Even our greatest scientists and philosophers won't ever be able to sit back with a comfortably complete understanding of reality. Not even close. And odds are, no matter how much some of us pine for a divine afterlife, it's neither probable nor scientifically likely we'll get one. I admit I could be wrong about that. I hope I'm wrong. I'd love to sit on a cloud eating chocolate and strumming a harp.

But to be pragmatic, *hope is a mistake*.

Which means most of us will have about eighty years to learn as much as we possibly can. About life. About the world. About the entire universe. If we're wise, we'll cram those eighty years like it's the night before a big exam. Because we're going to encounter a ton more questions than answers. And we're going to dream far more than we're able to *do*.

And then, at the end, our knowledge and wisdom will leave us.

Probably.

And that's ok.

Sort of.

If this is *it*, if our lives are to remain full of questions and theories and small discoveries followed by *more* questions, so be it. Maybe it really is all about the journey, and not all about how things begin or end.

Maybe.

Someday, a hundred-thousand generations from now, the accumulation of human knowledge and wisdom might reach a critical mass. We might grasp the nature of everything, and we might render books like this (and thousands of other texts) obsolete. Wondering might no longer be a thing. Questions might go extinct.

But it's not happening soon.

And certainly not during my lifetime.

In reading this book, one might be tempted to assume a certain arrogance on my part. It might appear as though I think I've got the right answers, and that I'm asking questions from an attempted position of know-it-all.

Rest assured.

I'm not.

My view is and always has been that I don't know *anything*.
I'm dead serious. Beliefs and opinions and theories are swell
things to have, I guess. It's human nature to possess them, to use
them in conversations, and sometimes even to employ them as
weapons.

But not for me.

Not ever.

I guess what I'm saying is: even though hope is a mistake ☺, I
hope the fact that you're reading this book means you subscribe to
a similar philosophy.

The philosophy of:

It's ok not to know.

It's ok to admit ignorance.

It's perfectly acceptable to ask questions for which there are
no knowable answers.

In fact, it's almost criminal not to.

Right?

Love,

J Edward Neill

Butterflies

The *Butterfly Effect* is defined as:

The phenomenon whereby a small change within a complex system can have large effects elsewhere.

In other words, a butterfly making the smallest alteration in the wind might set off a chain reaction causing a hurricane two weeks later on the opposite side of the world.

Or something like that.

Think of an important event that has taken place during your life or the life of someone you know.

Now think about how that event came to be.

What was the butterfly?

Does Your Head Hurt Yet?

Suppose that when you wake up tomorrow, you know everything.

As in *everything*.

Everything that has ever happened.

Everything that is currently happening.

And everything that will happen.

And you'll know the reasons *why*.

In theory, this would enable you to answer any possible question

the universe could pose to you.

Or would it?

If you knew everything there was to know, is it possible such

knowledge would only create *more* questions?

Meaning you could never truly know everything, because for each

answer you had, a new question would pop into

existence…infinitely.

Well?

Chills

Cryonics.

In other words, freezing terminally ill humans with the aim of restoring them in the future using medical advances that have yet to be invented.

For purposes of this question, let's assume someday medicine will reach a point of actually being able to thaw and restore a cryonically frozen person.

Now imagine you're terminally ill. Would you allow yourself to be cryonically frozen if the restoration process were guaranteed to be invented in:

50 years?

150 Years?

1,000 years?

10,000 years?

The Pilgrimage

On a long enough timeline (10,000 years or so) it's *possible* humanity might develop and utilize the technology to spread our species to other planets.

Meaning:

Interstellar travel

Terraforming

Planetary colonization

And given an even longer timeline (10,000,000 years or so) it's possible (however unlikely) humanity might be able to occupy every habitable planet in our galaxy, maybe even beyond.

Suppose it were *definitely* possible.

At what point do you believe humanity would be wise to stop its own spread?

Any chance you'd say *never*?

No Accidents?

Scientifically speaking, everything happens for a reason.

Literally.

Meaning on a physical level every interaction in the universe has a definitive physical cause, no matter how obscure.

And thus, universally speaking, there are no accidents.

No luck.

No true randomization.

But…

Things like human emotions, impulses, and ideas don't *necessarily* fit into any standard scientific construct.

Meaning the reasons *behind* several of humanity's physical actions aren't exactly known.

Meaning human activity can disrupt the universe's interactions.

So are there accidents after all?

The Doctrine of Double Effect

There's a runaway train.

Aboard it are five people.

Thing is: you're standing at a lever which, if you pull it, will switch the tracks and save the five people.

But…

If you pull the lever, you'll redirect the train to a portion of the tracks where a woman and her baby are standing, totally unaware of what's about to happen.

If you *don't* pull the lever, five people will die through your inaction.

If you *do* pull the lever, two people will die due to your direct action.

Which choice is correct?

Soul Food

First, a few definitions:

Soul - sōl/ (noun) - the spiritual or immaterial part of a human being or animal, regarded as *immortal*.

Mind - mīnd/ the component of a human being or animal that enables them to be aware of the world and their experiences.

Despite our many technological advances, one thing science has yet to define and/or locate proof for is the existence of souls.

Also hard to define are the processes of thoughts, feelings, and perceptions.

So which is more likely?

Souls and consciousness are physics-based, and will one day be completely explainable by science?

Or

Souls and consciousness operate under separate parameters and are not bound by any known laws of physics?

The Alpha and the Omega

Is it possible:

The only true purpose for all life everywhere is to survive and replicate?

The Fear of Consequence

Morality is a slippery concept.

Good and evil have shaky foundations.

Point is: what's considered good and noble in one culture isn't always viewed likewise in another culture.

Moreover, left without supervision, individuals tend to take a lot more liberties with morality. If the authorities aren't around, people will assault, loot, and murder more than if there's a police car nearby with its lights blazing.

Which begs the question:

Are people 'moral' only because they fear punishment if they're not?

If the concepts of authority and law didn't exist, and no punishment awaited disturbers of the peace, would the world gravitate toward violence and entropy?

If so, does that mean morality only exists beneath the fear of consequence?

The Paradox

Socrates, famed Greek philosopher, once posed that any man who believed himself wise was in fact *not* wise at all.

His reasoning was that any man who boasted of his own wisdom had no true awareness of his ignorance, and therefore could not be considered wise.

Meaning, paradoxically, Socrates might have been among the wisest men of his time, if only because he understood and accepted the vastness of his own ignorance.

The question:

Does being wise mean acknowledging one's own ignorance?

Taking up Space

The adage goes something like:

"We're a part of something larger than ourselves."

The implication is that humanity is a tiny, yet important cog in the universal machine.

But what if we're not?

Think about this: nearly *all* the physical places in the universe are inhospitable, if not downright hostile to living organisms. Not just to human life, but to all life. And not just a little hostile, but completely, utterly deadly.

The implication is that the universe might not appreciate life as much as *life* appreciates life.

So…

Are we truly a part of something larger than ourselves?

Or are we just taking up space?

Dictionary.You

Define the difference between *science* and *religion*.

The Wrath of _____?

In many religions, there exists a day of reckoning. A day of extreme destruction, cleansing, or a spontaneous revelation of a new age of consciousness.

The theological term for the study of these end-time events is known as *eschatology*.

But eschatology is a little tricky to pronounce, so for now we'll call it Judgment Day.

Now…

Given that in most religions, an imperfect humanity was created by a perfect divine being, is it *fair* for humanity to be judged, destroyed, or otherwise altered on a worldwide scale?

Wouldn't the more logical action be to judge the *creator* of our imperfect system, rather than its temporary occupants?

Worth

Considering the universe and all the people, places, and things within it, assign a tangible monetary value to the following people, and do so as objectively as you can.

Your child

An adult stranger on the street

Someone else's newborn baby

A murderer

Yourself

The value must be exact. It can be any value between $0.00 and $100,000,000,000,000 (that's one-hundred trillion.)

Worth More

Now let's up the ante.

For purposes of this exercise, we're making up a point scale.

We'll call it the *Universal Point System*.

The system goes from 0-10. A *zero* means no value at all to the universe. A *ten* means the highest possible universal value. A *five* is somewhere in the middle.

For each of the following, assign a value based on the Universal Point System:

A drop of water

The planet Mars

A single human being

A blue whale

Earth

A star

A galaxy

Gravity

Game of Thrones

Choose which king you would prefer to rule the kingdom in which you live:

The Sword: *Strong, decisive, militarily powerful, yet prone to rash judgment*

The Peacemaker: *Calm, benevolent, generous, yet unwilling to take risks*

The Lawyer: *Highly intelligent, even-handed, able to solve complex problems, yet entirely non-emotional*

The Philosopher: *Visionary, wise, deeply thoughtful, yet not at all concerned with tradition or religion*

And which king would you *least* want on the throne?

The Grand Illusion

Ultimately, we have no objective proof that anything we know or experience is real.

However improbable, it's possible that due to our limited scope of sensory input, we exist in an illusion or machine controlled by an entity who exists beyond our perception.

If this were true, would you want to know?

Really?

It's All Gravity's Fault

Modern science does a spectacular job of explaining the exquisitely precise laws of physics governing our universe. But...

There's no real way for anyone to know why our universe is governed by *these* laws as opposed to, say, a *completely different set* of physics laws. It's an unanswerable question.

In your mind, does this:

Indicate the intelligent design of our universe?

Mean we've still got a lot to learn scientifically?

Indicate a certain amount of randomness in terms of the universe's creation?

It's pointless to speculate on why we follow one set of physics laws instead of another?

(You can answer with more than one of these.)

The Value of a Thing

In many instances, science is an expensive habit.

Educating, experimenting, and exploring consumes resources and time.

In particular, space exploration costs *lots* of resources and time.

Some of these ventures yield amazing technological advances.

Others reveal key information about the universe we live in.

And still others try, but fail to accomplish much.

In your personal view, is…

The primary purpose of science to improve the quality of our lives and increase our level of convenience?

OR

The primary purpose of science to educate and enlighten us about the nature of our existence?

Are there any specific fields of science you would eliminate due to high cost and low yield?

Strung Together

In a nutshell, String Theory (aka: the Theory of Everything) goes something like:

All of the different *fundamental* particles of the universe are really just different manifestations of one basic object: *a string*. Not literally a piece of string, but a tiny, tiny object that moves similarly to one.

Meaning that if the string moves in one way, it's a *proton*. But if it moves in a different way, it's an *electron*.

In your mind, if this theory turns out to be true, does this mean the entire universe and everything in it is made of the same exact stuff?

Quotable

Which of the following do you believe?

"Words exist because of meaning. Once you've found the meaning, you can forget the words." ~ Chuang Tzu

"Man is the only animal who enjoys the consolation of believing in a next life. All other animals enjoy the consolation of not worrying about it." ~ Robert Brault

"The tighter you squeeze, the less you have."

"The opposite of a correct statement is a false statement. But the opposite of a profound truth may well be another profound truth."
~ Niels Bohr

"You have to do it by yourself. And you can't do it alone."
~ Martin Rutte

There is no *Slightly*

Chaos Theory goes a little something like this:

In complex systems (a machine, the Earth, or even the entire universe) having *slightly* different initial conditions will yield dramatically different outcomes.

For example; if you cloned Earth down to the last atom and started their existences at the exact same time, but made the temperature on Earth # 2 just 0.01% warmer, over the course of time the two planets would end looking very different.

Now…

Suppose you didn't mess around and made the two Earths *exactly* the same, including the starting temperature, would they look the same 1,000 years later?

Obviously the answer is unknowable.

But what do you *think*?

Nothing is Not

Nothing is less than what most people think it is.

Nothing isn't an abyss. An abyss is *something*.

Nothing isn't a vast empty space. A vast empty space is *something*.

Nothing is *nothing*. No space. No emptiness. And most importantly, no potential to ever be anything more than nothing.

Now comes the question of belief versus truth.

Why is there *something* instead of nothing? Why does our inconceivably huge universe exist instead of nothing? Why are there stars, planets, human beings, paper clips, and Pokemon…instead of nothing?

If your answer is anything other than, "I don't know," you have some serious explaining to do.

Deterministic Dilemma

Fact: atoms behave in a probabilistic manner (meaning they *usually* do *exactly* what they're predicted to do.)

Fact: our brains are made up entirely of atoms

Fact: recent scientific advances using Functional Magnetic Resonance Imaging (fMRI's) have determined that our brains typically make decisions several seconds *before* we're even aware of it

So...

If our brains, being made of things that do exactly what they're supposed to do, and being capable of deciding things before their hosts (us) are aware of it...

...does free will exist?

No Time for an Investigation

Aside from discussing the mundane, is it arrogance when someone says they *know* the truth beyond a shadow of a doubt?

I'm Just Here for the Free Drinks

When someone asks, *"What is the meaning of life?"* perhaps they should be a bit more specific.

Because…really…there are multiple conditions to such a question.

Are they asking, *"What is the meaning of ALL our lives?"* which indicates a grand, unified purpose behind humanity's existence.

Or maybe they're trying to ask, *"What is the meaning of YOUR life?"* which implies our purposes are self-created.

Or, given that our individual views on the subject are extremely subjective, maybe all they're really asking is,

"Is there any meaning to ANYONE's life?"

So the question is: which one of these questions is the right one to ask?

Is the meaning of life self-generated, as in it's different for each individual living thing?

or

Do all lives have similar meaning?

Collisions

Bill and Sara have just had an argument.
The day after the argument, Sara calls Bill to tell him he really hurt her feelings. She says she's upset, and will be for a while.
Bill rejects this notion. He claims steadfastly that he didn't hurt Sara's feelings, and states she has no reason to be upset.
This is clearly a collision of perception.
Both Sara and Bill firmly believe their positions.

Objectively speaking, did Bill actually hurt Sara's feelings?

Explain your answer.

I'm Done with this Test

Nirvana

It literally means: *blown out (like a candle) or 'life extinguished.'*

It's not meant as a negative thing. It's meant to represent a profound peace of mind, an utter release from egotism, and an end to the cycle of karma.

In technical terms, Nirvana is specific to the Hindu faith. It's a state of consciousness ending the cycle of reincarnation.

But let's move beyond that.

Let's suppose Nirvana is a worthwhile state of mind to reach for *anyone* with *any* belief system. Utter peace of mind sounds pretty ideal, right?

Imagine Nirvana is a legitimate thing, and that you have but one single lifetime (your life, right now) in which to reach it.

Would you try to achieve it?

And how would you go about doing so?

You are (*only*) what you Eat

Materialism is a doctrine stating that all things in the universe, without exception, are bound by the laws of physics.

In other words, there is matter, and there are forces that make matter do stuff, but there are no meta-forces guiding the matter.

Also, materialism disallows separation of consciousness from all other physical interactions.

Meaning our brains and our thoughts are *one*.

Now, the argument starter:

If materialism is valid, does it…

A. *Imply there can be no intelligent creation?*

B. *Mean free will does not exist?*

C. *Imply a certain pointlessness to our lives?*

D. *Eliminate the need for morality?*

E. *Not bother you in the slightest?*

(Choose as many as you like.)

Manifest Reality

Conversely to materialism, we have *idealism*.

Idealism asserts that our entire existence is a mental, consciousness-driven one, and therefore is utterly mutable by our perception of reality.

While many branches of idealism recognize the existence of physical matter, nearly all of them declare that our reality is *conceived* by our consciousness.

Now, the argument continues:

If idealism is valid, does it...

A. *Mean Santa Claus exists if a child truly believes in him?*

B. *Imply the likelihood of intelligent design?*

C. *Allow each individual his or her own self-constructed reality?*

D. *Allow for events beyond those explained by physics, including supernatural and spiritual occurrences?*

E. *Mean individuals' separate realities are constantly crashing into each other?*

Why Bother?

There exist many humans in the world who maintain a firm belief

in determinism.

In fate.

In destiny.

In other words, they feel that no matter what actions humanity

takes, the outcome of our existence will be the same.

Why then would anyone who believes this philosophy:

Look when crossing the road?

Go to work?

Eat?

Ever do anything?

Or were they pre-determined to do or not do these things?

A Party for the Gods?

If in fact there was a Big Bang, meaning the almost impossibly powerful explosion that birthed our universe…

…what happened before it?

Or maybe there was no *before*?

Born Yesterday

It's known that extremely high velocities slow time relative to
objects not moving as fast.
Meaning that for an object traveling at high speed through space,
time will move slower than for an object sitting on Earth.
With this in mind, is it possible that the light from our universe's
earliest stars (*which travels at 671,000,000 mph*) was only just
created *moments ago*?

Think about it.

Changeling

Every moment of your waking life, you are *experiencing*.

Witnessing.

Learning.

Forgetting.

Even if you're not aware of it, it's happening to you right now.

Therefore, scientifically speaking, your brain isn't in the same physical state it was at the beginning of this sentence.

So...

If your state of mind is ever-changing, and if your exact experience of life is always different than it was just moments ago...

Are you a different person now than you were thirty seconds ago?

What Will Be, Will Be

Imagine a model of the universe in which there is *definitely* a god.

For many, this won't require any imagining at all.

Now…

In most religions, the god or gods are not only all-powerful, but also all-knowing.

Meaning they know what *has* happened, what *is* happening, and what *will* happen.

And scientifically speaking, if the future is a known absolute, then by its very nature it cannot be changed.

Meaning everything humanity does is fixed.

Fated.

Predetermined.

Therefore, if in this model all actions are predetermined, why should any human be held responsible for any action ever?

Including the horrible ones…

We Can But Try

Objective - /əbˈjektiv/ -

A person who is not influenced by personal feelings or opinions when considering and representing facts.

In other words, completely unbiased and unmoved by anything but the truth.

Can a person ever be 100% objective?

If so, can they do it *always* for *everything*?

Inconceivable

The universe

Life

Gravity

Consciousness

These all have something in common:

They're all parts of our reality, but also have an unknowable

source of creation.

In other words, we don't know *why* these things exist.

Lacking complete data on the origin of these (and many more

things) does humanity have little choice other than to declare these

phenomena are *real* and move on?

Or is it our scientific duty to pursue a reason why these things

exist?

Play the Percentages

How much of your total knowledge is *firsthand*?

As in, *you* saw it. *You* experimented with it. *You* understand it due to *your* direct contact with it.

Examples:

You threw a ball into the air and watched it fall down.

Thus...gravity

You hit your thumb with a hammer.

Thus...force and pain

You observed a tree react to the climate for an entire year.

Thus...seasons

And how much of your knowledge is either derived from inference or hearsay? As in, someone else told you about an event, you saw the event on TV, or you read about it in a magazine?

Give percentages for your direct and indirect knowledge.

Bonus question: is a piece of knowledge gained indirectly nothing more than a *belief* until you witness it or prove it firsthand?

Turing

The Turing Test - designed by Alan Turing, an assessment of a computer or artificial intelligence performed by a human with the intent of gauging whether or not the computer or AI possesses consciousness.

Got that?

Now let's go beyond computers.

Suppose you performed this test *not* on an attempted AI or a computer, but on *another human*.

Now, two crazy questions:

Is there anyone you've met in your life who would fail this test, thus failing to convince you they had consciousness?

And…

Can consciousness be reproduced? And if not, does that prove the existence of souls?

The Conjecture Clock

First, here's a few interesting measurements of time:

Attosecond – Currently the smallest division of time. Approx 10^{-18} seconds.

Megasecond – Approx 11.6 days

Galactic Year – The time it takes for the Sun to orbit once around the Milky Way's center. Approx 230 million years.

Exasecond – Approx 31.7×10^9 years. (more than twice the age of the universe.)

Now, the real question:

Does time exist?

Or is it simply a human construct?

When answering, *take your time.*

Thought Police

Is it immoral to be willfully ignorant?

Two Worlds

Humans experience some pretty strange phenomena.

Things like *déjà vu, synchronicity, placebo effects.*

Despite the belief (or hope) that these effects are spiritual or otherwise outside the realm of explanation, most of them have causes rooted in science.

And yet...

A few phenomena exist that have yet to be fully explained.

Things like *ghosts, past-life memory, ESP.*

Which leaves us with three distinct possibilities:

These things don't really exist. People make them up.

These things do exist, but have scientific reasons we've yet to find.

These things do exist, but have causes outside the realm of science.

Which one do you think is most likely?

And why?

Plato's Play-Doh

Plato, famed philosopher, once suggested that a bad democracy is *better* than a bad tyrant.

Mostly because in a bad democracy *everyone* is participating in the badness.

Think hard.

Would you prefer a lone, bad tyrant or a bad, malfunctioning democracy?

Or are both equally horrible?

The Sun will Rise Tomorrow. Won't It?

If you can, name three things or phenomena it's acceptable to *believe* in without having actual objective proof of that thing or phenomenon's existence.

It's *Your* Turn

"Where must we go, we who wander this wasteland, in search of our better selves."

No, it's not a quote from Socrates, Plato, or Neil deGrasse Tyson.

It's from George Miller, Hollywood director.

But it's still a good question.

Specifically, it's similar to the author of this book's big life question:

Where must we go and what must we do to find our better selves?

Now it's your turn.

The universe has agreed to answer one single question of your choosing.

What will your question be?

The Wellspring

Must all knowledge be derived from using our actual senses to learn it?

Meaning, must we be able to see, hear, touch, taste, or smell something in order to know *without doubt* something is real and true?

Examples:

We touch a baseball and count its seams - therefore the ball is real

We feel a cold wind blowing over us - therefore the wind is real

We see other people walking by - therefore they are real

Or...

Can *some* knowledge, such as the existence of a god or the presence of a ghost, be assumed without sensory proof?

Explain.

Need to Know Basis

Buddhism

It's a philosophy in which the primary purpose of life is to earn freedom from restlessness, karma, and unease.

One method some Buddhists seek peace is to avoid questions that focus on non-physical matters. They believe a focus on the *metaphysical* aspects of life only leads to deeper speculation, thus more questions, thus greater *unease*.

Unease which is directly oppositional to their peace-of-mind goal.

Would most of humanity find happiness easier to achieve if they abandoned the metaphysical questions they have about life and the universe?

Yesterday versus Tomorrow

Regret

Worry

Two powerful states of mind that have long influenced humanity.
Regret arises from past events, which, barring the invention of time travel, are unchangeable.
Worry stems from fear of what the future may bring. The future may or may not be changeable, depending on your view of determinism, but it doesn't appear to stop people from worrying.

So...
Which one is a greater influence on humanity?

Or are they equal?

Apollo and Dionysus

In Greek mythology:

Apollo represents reason and rationality.

While *Dionysus* is the god of chaos and irrationality.

This dichotomy may have originated in myth, and yet like most

Greek stories, it has its roots in human philosophy.

The point:

Humans are capable of great reason and rational thought, and yet

are just as often inspired to deeds of great passion, some

constructive and some perhaps not so constructive.

The questions:

Does this duality accurately describe humanity?

Are we split into two powerful spheres, reason and emotion?

Or are these things not truly separate, and actually just different

expressions generated from a single source: our brains?

The Duplicant

Scientists have begun the study of mapping the human brain and 'creating' memories via experimentation using the portion of our brain known as the hippocampus.

Also...

Studies have been conducted to test the viability of *storing* human memories indefinitely.

Using such methods, it's possible one day we'll be able to erase the memories we don't want, store the ones we need for later, and perhaps even create memories of things that never happened.

Now suppose you were in a car accident.

You were rendered comatose for five years.

You were declared *brain-dead*.

But...before the accident, you had all your memories stored in a computer. And upon waking in a stupor from your coma, these memories were successfully re-installed into your brain.

Is the person waking up from the coma *you*? Or *not* you?

Promethean

Does humanity possess a mandate to continue our technological development until its utter end?

Meaning, should we advance ourselves technologically as far and as fast as possible?

If not, at what point during our advancement should we consider slowing down or stopping?

No Place Like Home

Reincarnation

It's a prime tenet of the Hindu religion, and is referred to peripherally by several other faiths.

The idea:

Every human and animal has a soul.

Upon death of the body, the soul can inhabit a new body, thus beginning life anew.

Let's imagine for a moment reincarnation is absolute fact.

Now suppose, due to some universal cataclysm, all life were permanently extinguished. It's entirely possible, even if it happens many, many years from today.

If all the souls in existence were suddenly rendered homeless (no bodies to inhabit) would they simply wander the universe for eternity?

And would that make the entire reincarnation concept pointless?

In the words of Pyrrho

Is it possible humanity can never really know the truth of *anything*?
As in, due to our limited scope of sensory input and highly human-centric perspective, can we only ever know the *appearance* of things, and never the reality of them?

It is NOT what it is.

"It is what it is."

You've probably heard this declaration a lot. And if you haven't, you probably will.

But scientifically, this statement is probably always wrong.

Earlier in this book, we discussed how the human mind is never in the same physical state as it was moments ago.

Now let's apply this to the entire universe.

Given that every piece of matter, from tiny atomic particles all the way up to stars and galaxies, is actually moving, shifting, and oscillating at all times, nothing we perceive is ever the same twice.

Electrons, cells, rocks, trees, rivers, asteroids, people…

Always moving at all times, even if imperceptible to the naked eye.

Does this mean you and everything you know is never the same as it was a millisecond ago?

Is everything *not* what it is?

Deus Ex

Imagine at some point in the future humanity successfully creates
AI.
Meaning we design an artificial intelligence that demonstrates a
clear ability for consciousness.

If we create such a thing, are we gods?

Not Quite What You Were Hoping

There exist numerous theories regarding the meaning of life.

Some predict a divine afterlife.

Others believe in infinite recycling of our souls.

Some believe in very specific versions of heaven and hell.

And still others say there's no meaning at all.

Everyone is guessing.

No one really knows.

Even so, the most common perception is that *if* there is a meaning,

it's probably a positive or at worst a neutral one.

But…

What if humanity one day learned our purpose is nefarious?

That perhaps humanity (or even all life) was engineered for a

negative purpose?

Is it possible?

If you learned such a thing were the truth, what would you do?

Finally, a Simple Question

Given the choice, would you rather know *how* the universe works, meaning you'd understand all the hard science behind each and every interaction taking place in our existence?

or

Would you prefer to know *why* our universe and all the individual objects within it exist, meaning you'd grasp the *purpose* behind everything?

Explain your reasons.

Something in our DNA

Are humans nothing more than highly intelligent animals?

Or are we something special?

Something unique?

Are we a species deserving a larger share of the world, maybe even the universe?

Climbing Up the Chain

The science behind humanity's physical evolution is well-known.

We've unlocked the human genome.

We've studied macro and micro evolution of our species.

Generally speaking, we've come to understand how and why our skeletons, organs, and brains are the way they are.

And then there's *conscious evolution.*

It's a theory asserting that due to our technology and understanding of the human condition, we now have the ability to control our future evolution.

Meaning we can either cooperate to improve our bodies, minds, and societies…

…or compete and ultimately destroy ourselves.

Imagine tomorrow the entire world agreed to *cooperate.*

What should be our first goal in the next step of human evolution?

Cartesian

"If you would be a real seeker after truth, it is necessary that at least once in your life you doubt, as far as possible, all things." ~
Rene Descartes

It's a pretty powerful statement.

Do you agree with it?

Why or why not?

Odds are: We're *all* Wrong

At the time of this book being published, the human population of Earth exceeds seven billion.

And among these seven billion can be found *thousands* of varying viewpoints, including:

Different religions

Different views of morality

Different social and political beliefs

Scientifically speaking, if *one* particular combination of beliefs happens to be absolutely true, then by rule the other belief systems *cannot* be true.

Meaning a huge percentage of the world's population holds a belief system that is likely incompatible with reality.

So…

Are you among the small percentage of people whose beliefs are true?

If so, does this mean everyone who disagrees with you is wrong?

That's *MY* Throne

Typically when discussing humanity's effect on Earth and all the lives and ecosystems on it, we tend to talk in the negative.

We use words like:

Wasteful

Dirty

Greedy

Destructive

Moreover, we tend to emphasize that humans have these qualities, and that other animals (often our victims) do not.

And yet, one must wonder...

If other species had the same tools we humans have (large brains, opposable thumbs, language, et cetera) is it likely they would take our place as Earth's dominant life form...

...and be just as destructive as we are?

The Monk

Is it possible for a human being to live without *believing* anything?

Meaning this person would have:

No opinions

No claims to knowledge without hard physical evidence

No spirituality

No philosophy beyond the material world

No religion

If someone could pull this off, would it be admirable?

Or is this an impossible state of mind for a human to achieve?

The Human Lens

Sixth senses aside, *everything* you know about the world, you know through the subjective lens of your human brain.

Meaning you only truly know what you see, hear, smell, touch, and taste.

You'll never know what it's like to see the world in the same way a cat does, or a bird, or a whale, or a bacterium.

Meaning you'll only ever experience the universe from a human point of view.

And more specifically, *your* human point of view.

So...

Does this mean your experience of reality is unique, almost isolated in its *filtered-through-a-human-lens* nature?

Or does this mean that physical reality itself is different for every single living thing?

If it Hasn't Happened by Now…

If, in the entire history of humanity, no one has yet been visited by someone from the future, does that mean time travel won't ever be invented?

Cardinals

The old saying goes:

"Do unto others as you would have them do unto you."

It sounds great, right?

Until you realize that many people wouldn't mind having some
pretty strange and awful things done to themselves.
Which, in theory, gives them license to do those same things to
someone else.
Now let's break this cardinal rule.
If you can, name at least one thing you'd like done to yourself that
would invalidate the *'do unto others'* adage.

And is there any such thing as a perfect ethical system?

I Am.

Human self-identity is largely a function of perception.

Meaning how we see ourselves appears to dictate who we are.

Or at least, who we *think* we are.

Moreover, if someone perceives themselves as being something (smart, stupid, strong, weak, et cetera) the tendency is for them to gravitate toward becoming that thing.

In their own eyes. And possibly in the eyes of others.

Now suppose you took two people and switched their minds into each other's bodies.

Meaning Jim's mind goes into Sara's body, and vise versa.

A year goes by after the switch.

Jim and Sara are now accustomed to being in their new bodies.

But are they still Jim and Sara?

Or, due to having completely different perceptions based on their new bodies, are they different people entirely?

If Time Were a River

We've already discussed time travel.

And the likelihood that it's impossible.

But…

Suppose it were possible.

Imagine you had the capability to enter a time machine and go back one-hundred and fifty years, roughly to the same time period during which your great-great-great-great grandfather lived.

If you were to kill your great-great-great-great grandfather, you would in effect prevent all his offspring from existing, including yourself.

So would that mean you couldn't have time traveled in the first place?

Leaving

Just a few of the potentially lethal obstacles preventing easy space travel are:

Time

Cosmic radiation (small, dangerous particles traveling at high speed)

Space debris (traveling at high speed)

Life-support

Psychological effects of prolonged time in deep space

Exiting and entering planetary atmospheres

Given these, and given the vast amount of resources needed to even *attempt* to conquer these, answer this one question:

Is deep space travel a worthwhile human endeavor?

Just Because

Set your existing religious beliefs aside for a moment.

Imagine, no matter what you normally believe, there is definitely *one singular divine being* responsible for the creation of all things.

A god, a goddess, or whatever.

Now…

Justify why this divine being created:

Viruses

Cancer

Black holes

Serial killers

Good luck.

Prior to recent scientific research, it was believed that every cell in the human body replaced itself approximately every 7-10 years.

This is now known to be false.

The neurons in the human brain are not replaced during our entire lifespan.

Even so…

Most of our other tissues are replaced, including fat, muscle, and bone cells.

Meaning the vast majority of your body is not the same body you had 10 years ago.

Does this mean you are a different person than you were a decade ago?

Or does this mean humans are in fact the contents of their minds, and not their bodies?

The Ache

It has been said that most of humanity suffers from a permanent
type of pain.
Not a physical pain, mind you, but a subtle unease and
dissatisfaction due to our mortality, our unmet desires, and the
transient nature of all life's experiences. (i.e., all good things must
come to an end.)

Do you feel it?
The quiet ache beneath your heart?
However small or large it might be?

Is it possible to cure?

Izms

From the following, choose which one(s) you associate with your personal philosophy of life:

Cynicism - The purpose of life is to live with virtue and in harmony with nature (not what you thought it meant, is it? ☺)

Agnosticism – Humanity knows nothing beyond that which it can touch

Pragmatism – The most valuable things are tangible and practical

Hedonism – Life's purpose is to pursue pleasure

Capitalism – Life's purpose is accumulate wealth for the benefit of yourself and your family

Theocentrism – God is a central fact of our existence

Nihilism - Life is without objective meaning, purpose, or value

Existentialism – The universe is unknowable, yet humans still have individual purpose and responsibility

A Shallow Grave

Famed scientist Stephen Hawking twice pronounced,
"Philosophy is dead."
His assertion is that the pursuits of philosophy, in particular
physics, have not maintained pace with modern science.

Given the rapid advance of science, technology, and our ever
increasing understanding of the universe's makeup, is traditional
philosophy still a worthwhile pursuit?

Or should we focus all our questions on hard science?

Holy ____!

Objectively speaking, it's logical to say that the most honest answer to many of life's deep questions is: "I don't know."

Is there life after death? – I don't know.
Are there aliens out there? – I don't know.
Is there a god? – I don't know.

In each case, *I don't know* is a responsible reply.
And by this reasoning, both atheists and devout believers appear presumptive in their beliefs about the workings and creation of our universe.
Perhaps only the agnostic point of view is truly honest.
That said, for each of the above three questions, what do you think is the *most likely* answer? And why?

Disaster Porn

Y2K

The end of the Mayan Calendar

Nostradamus

Polar ice caps melting

The theme here is obvious. For hundreds of years, scholars, theologists, and even scientists have attempted to predict the end of the world.

None have been correct. Yet.

Regardless, many humans appear concerned with how, when, and why Doomsday will arrive.

Perhaps more important than trying to guess when the world will end is understanding the psychology behind humanity's morbid obsession with it.

The question: does human interest in the idea of Doomsday reflect a certain unhappiness with the world?

I ♡ Sagittarians

Astronomy – the science of celestial objects, space, and the physical universe

Astrology - the study of the movements and relative positions of celestial bodies interpreted as having an influence on human affairs and the natural world

So…

You probably know:

…the stars contained in constellations are often millions of light years apart, having no real relation to one another

…planets and other celestial bodies have no known personality traits

Even so, several modern societies practice and follow *astrology*, identifying themselves in relation to constellations, attributing moods and life events to the movements of planetary bodies.

Why do you think this is?

The Vat

Imagine you and I and every other person in the world aren't really alive.

We're just constructs in a system created by beings (or machines) who exist beyond our collective consciousness.

They're watching us. Controlling what we experience. Setting the parameters for all the physics we hold dear.

Don't laugh.

It's entirely possible, even if improbable.

The question is:

If tomorrow you became aware of this fact, meaning you saw a glitch in the system and became aware of humanity's condition, would it matter?

To you personally?

To anyone?

Lego Babies

Within the next hundred years (likely much sooner) it's probable that sufficient technology will exist for *all* parents to decide *every* aspect of their child's genetic disposition.

This means parents will be able to decide and/or edit their child's:

Sex

Skin tone

Height

Weight

And much more.

Suppose you're having a kid and this technology is readily available.

Do you use it? And to what extent?

Right or Might

Is it better to:

Do the right thing and fail?

or

Do the wrong thing and succeed?

(Note: to answer this question, you'll have to assume that such things as right and wrong actually exist.)

Think Twice

French author André Gide once posed:

"Believe those who are seeking the truth; doubt those who find it."

Do you agree with this statement?

Have you ever believed you knew a significant truth, only to find out later you were wrong?

And are you more objective as a result of that experience?

Context is Everything

Thomas Jefferson's famous quote goes likewise:

"...that all men are created equal, that they are endowed by their creator with certain unalienable rights, that among these are life, liberty and the pursuit of happiness."

But are all humans truly created *equal*?

It's probable Jefferson didn't mean to imply that everyone is equal physically, but rather that the inherent value of a human being is the same no matter their origin, belief system, or life circumstance.

But again, are *all* humans equal?

Is a cruel, vicious, murdering war-criminal as valuable as a kindly philanthropist?

Is a brain-dead person, though still human in physical terms, as valuable as a healthy, adorable infant?

Or is the value of a single human life completely subjective, and utterly different depending on perspective, therefore negating the concept of equality?

Magnetic Personalities

Describe whether or not you agree with the following two sentences:

Some people are drawn closer to one another by a nameless magnetism.

Others are divided by an unreasonable dislike.

It's not very scientific, given that humans aren't known to be particularly magnetic.
So if you believe both sentences are true, explain why.

Sartre's Sabre

Given that the answer to most of life's questions (and indeed most of the questions in this book) are, *"I don't know,"*

...is it fair to say:

Since we have no evidence of the correct answers to life's great mysteries, are we free to create our own?

Ascetic

Wine

Liquor

Sex

Fine foods

Drugs

Just some of the things humanity indulges in.

For the sake of this question, let us assume none of these things are *bad* or *immoral* by themselves. While it's possible the people overindulging in them might do harmful acts, the actual wine, sex, food, et cetera aren't to blame.

That said, is a person who indulges in *none* of these a stronger person morally than someone who indulges in them often?

Does denial (or severe limiting) of one's indulgences make a person better?

Or do indulgences have no bearing on a person's *goodness*?

Embracing the Shadow

Suppose the following three statements were true beyond any
shadow of a doubt:

No gods or goddesses exist

After death, there is only nothingness

Humanity is inconsequential to the universe

Given these statements, what incentive would a human have to be
good instead of evil?

And…

Would good and evil even be relevant?

Sub-Objective

Are you what you *are*?

Meaning, are you flesh, bones, and synapses firing? An animal through and through?

Or are you what you *think* you are?

Meaning, are you the sum of your experiences, your instincts, and your thoughts? A temporary inhabitant of a borrowed body?

Or are you both?

Like Father, Like Son

In most religions, gods and goddesses are depicted as having
human or animal characteristics.
Many times these physical traits are exaggerated. But most often
they resemble *something* that appears on Earth (people, elephants,
serpents, et cetera.)
For the sake of this question, *no matter your religion*, suppose
gods and goddesses are absolutely real.

Do you believe a god presents itself as a familiar earthlike thing in
order to make you (their worshipper) feel more comfortable?

Or is it more likely, assuming gods exist, that all the statues and
pictures made by humans probably don't resemble what actual
divine beings look like at all?

End of Ages

Is it possible (or even probable) that ages from now, much of the science and philosophy we now take for truth will be revealed as false, and a newer, truer system of knowledge be put into place? In other words, could it be a lot of the things we *think* we know are completely wrong?

Also…

Is it possible (or even probable) that the only period of time during which humanity will know the truth of everything (or close to everything) will be mere moments before the end of our existence?

Can I Get Your Number?

Almost everything in the universe can be discussed in terms of
numbers.

Atoms

Compounds

Chemical interactions

Temperatures

Velocities

Forces

Mass

Time

All of these things can be discussed in utter detail using the *same
number system*. Meaning, numbers are a unifying language for
much of science.

Does this mean numbers are real?

As in actual physical things, and not just a concept?

Choose Your Own Adventure

From the following, choose which one you *hope* is what happens after your death:

- *People who exhibit sufficient good in life go to a heaven of some sort, while everyone else suffers a worse fate*

- *When we die, all that we are is forever lost*

- *Reincarnation; either as a human again or a different animal type*

- *We ascend to some higher form of consciousness, meaning we're no longer human, but we retain some of what we once were*

- *We roam as spirits either forever or for a period of time*

And now choose which one you believe is *most probably* the truth.

Nihil

We've talked about nihilism before.

Nihilism - Life is without objective meaning, purpose, or value

But can someone truly be a complete nihilist?

Wouldn't someone with such a worldview immediately commit suicide, being utterly hopeless as they are?

For that matter, considering the questioning nature of the human mind, can anyone believe in any '*ism*' 100% of the time, no doubting ever?

A Big Limb to Go Out On

For purposes of this question, let us suppose that morality indeed exists. That good and evil are definable things. And that humanity has a tendency, though not a mandate, to be good instead of bad. With that in mind, are any of the following reflective of a tendency toward immorality?

Believing in something without objective proof

Teaching children to believe in things without objective proof

Denouncing other humans for their beliefs without objectively disproving those beliefs

Desiring segregation of people based on their beliefs

Desiring the conversion of other people to the belief system of your choosing

When We Were Evil

Throughout most (though not *all*) of the modern civilized world, codes of morality have been put into place that are largely in agreement with each other.

Generally speaking, assault, theft, rape, and murder are among the actions viewed as immoral, and thus are illegal.

And yet...

Portions of societies have existed (and still do exist) in which these actions *aren't* viewed as immoral. In other words, they're perfectly acceptable to do in certain contexts.

Now imagine the entire world follows this reverse code.

Assault, theft, rap, and murder are allowable.

Will human nature, believed by many to be generally good, eventually overcome this system?

Or are humans capable of remaining 'evil' indefinitely?

Product of the Past

Since we are, *none* of us, responsible for our own presence in this world, meaning that none of us created ourselves or willed ourselves into existence, does that reduce any of our personal responsibility in this life?

In other words, *every* human alive was given life without his or her consent. We didn't ask for this particular existence, and in fact, if given a choice, many humans might have chosen a different existence altogether.

Does not having chosen this life allow for a certain *moral flexibility*?

Or...

Must we accept a moral responsibility whether or not we asked for it?

And if so, why?

It Makes For a Great Party Favor

What role does *love* play in the universe?

Meaning, is it a chemically-induced emotional state designed to help living things procreate, protect our younglings, and generally not slaughter each other whenever we feel like it?

If you believe it's deeper than that, and that it has some meaning beyond chemical compounds floating through our blood, here's your chance to explain.

Seriously.

Go for it.

Forsaken

In the U.S., approximately 63% of people maintain a firm belief in a god. And of the remaining unaffiliated people, 83% maintain a general belief in a divine creator.

In many countries worldwide, the numbers are much higher.

That's a *lot* of people. In fact, odds are *you* are one of the believers, whether firm or not so firm.

Now imagine tomorrow every single human on Earth were struck with undeniable, absolute evidence that there was in fact *no god at all*.

Consider that many cultures rely on the existence of a god or gods for their moral codes, traditions, and indeed, their very happiness.

With this belief shattered, would these cultures descend into chaos? Would morality go right out the window?

Or would most people reject the proof and continue to believe as they wish?

Everything for Nothing?

First, a definition:

Determinism - a theory stating that acts of free will, natural occurrences, and all social events are determined by preceding events or natural laws

Meaning that no matter what we do, how we do it, or when it happens, *determinism* states the end result will be the same.

Kinda morbid, right?

Now…

Considering the universe and all the vast intergalactic events happening out there in the void, do you believe anything humanity does now or in the future will ultimately make a difference in how the universe moves along with its existence?

If not, meaning if humanity makes no difference in the end, answer this:

What is our value?

On Your Knees, Protons

In terms of scientists, atheists, and non-religious people...

...is physics god?

Moribund

If physical death is the absolute end of a human's existence,

And if the Earth is destroyed before humanity is able to populate other worlds,

And if in a future state the universe contracts onto itself or renders itself otherwise inhospitable to all life,

...is there still genuine value in the human experience?

Negating This Book

Define the value (if any) in asking questions to which the most honest answer is, *"I don't know."*

The universe, vast and improbable, appears to follow its self-made laws throughout.

Except when it doesn't.

What matters is that we have no idea what matters.

Perhaps one day humanity will draw back the curtain and figure out the science behind the science. Maybe we'll reach the how and the why of everything. Maybe.

Until then we're just here to exist. And observe. And theorize.

And more than anything, ask questions.

A thousand years from today, nearly all of humanity is jacked-In.

We sleep, connected to machines, dreaming our lives away.

J Edward Neill's…

J Edward's Novels:

Eaters of the Light trilogy

Darkness Between the Stars

Shadow of Forever

Eaters of the Light

Tyrants of the Dead

Down the Dark Path

Dark Moon Daughter

Nether Kingdom

A Door Never Dreamed Of

Hollow Empire – Night of Knives

The Hecatomb

Coffee Table Philosophy:

Reality is Best Served with Red Wine

Life & Dark Liquor

The Small Book of BIG Questions

444 Questions for the Universe

101 Questions for Single People

101 Questions for Couples

The Ultimate Get to Know Someone Quiz

Big Shiny Red Buttons

101 Sex Questions

About the Author

J Edward Neill writes dark fiction, sci-fi, horror, and philosophy – all for adult audiences. He lives in North Georgia, where the summers are volcanic and winters don't exist. He has an extensive sword collection, a deep love of wine and scotch, and a chubby grey cat named Noodle.

He's really just a ghost.

He's only here to haunt the earth for few more decades.

Shamble after J Edward on his websites:

TesseraGuild.com

DownTheDarkPath.com

Téssera